PENGUIN BOOKS

WEATHER FORECASTING: THE COUNTRY WAY

Robin Page was born in 1943 on the small Cambridgeshire farm where he still lives. He is author of over thirty books, including *The Decline of an English Village*, *The Wildlife of the Royal Estates*, *The Hunting Gene*, *The Great British Butterfly Safari* and *The Revolting Peasant*. He is founder of The Countryside Restoration Trust and presented *One Man and His Dog* for several years.

Robin Page

Weather Forecasting: The Country Way

With wood engravings by
Thomas Bewick

PENGUIN BOOKS

PENGUIN BOOKS

Published by the Penguin Group
Penguin Books Ltd, 80 Strand, London WC2R 0RL, England
Penguin Group (USA) Inc., 375 Hudson Street, New York, New York 10014, USA
Penguin Group (Canada), 90 Eglinton Avenue East, Suite 700, Toronto, Ontario, Canada M4P 2Y3
(a division of Pearson Penguin Canada Inc.)
Penguin Ireland, 25 St Stephen's Green, Dublin 2, Ireland (a division of Penguin Books Ltd)
Penguin Group (Australia), 250 Camberwell Road,
Camberwell, Victoria 3124, Australia (a division of Pearson Australia Group Pty Ltd)
Penguin Books India Pvt Ltd, 11 Community Centre, Panchsheel Park, New Delhi – 110 017, India
Penguin Group (NZ), 67 Apollo Drive, Rosedale, Auckland 0632, New Zealand
(a division of Pearson New Zealand Ltd)
Penguin Books (South Africa) (Pty) Ltd, 24 Sturdee Avenue,
Rosebank, Johannesburg 2196, South Africa

Penguin Books Ltd, Registered Offices: 80 Strand, London WC2R 0RL, England

www.penguin.com

First published by Davis-Poynter Ltd 1977
Published in Penguin Books 1981
Reissued in this edition 2011
1

Printed in England by Clays Ltd, St Ives plc

ISBN: 978-0-241-95306-8

www.greenpenguin.co.uk

Contents

Weather Forecasting: The Country Way

Introduction

Living on a farm, as I do, one of the most vital programmes that I see on the television is the weather forecast. Ploughing, drilling, haytime and harvest all depend on the weather, and so an accurate forecast is of great importance. Yet it is strange to record that as the weather forecasting service has grown in size and expense, so its predictions seem to have become more inaccurate.

I therefore rely almost entirely on the old handed-down wisdom of country weather lore, and indeed for many years it has played an important part in life on the farm. My farming father has always claimed that barley should never be sown in the spring until 'the soil feels warm to your bare backside', and although he has never tested the soil that way himself, as far as I know, it means wait until the land warms up with the improved weather of spring.

Red sky at night, shepherd's delight,
Red sky in the morning, shepherd's warning

is the most well-known piece of weather lore, and for many years we have taken more notice of that than details of 'occluded fronts' and 'high pressure over the Azores'. Some people are sceptical about the simple rhymes, verses, and tales of odd animal behaviour which are said to forecast weather. Yet, recently, when I was with a gamekeeper on a warm clear day, he suddenly saw a rabbit wandering about at the wrong time of the day and immediately he forecast violent storms. I thought that he was overplaying the role of 'backwoodsman', but within twenty-four hours there were gale-force winds, torrential rain, and widespread flooding. Shortly afterwards, during the storm, a Devonian housewife put her cutlery away wet, to prevent the stainless steel from attracting the lightning into the house. Although weather lore is considered to be 'old fashioned', much is still in everyday use and most of it is based on simple but accurate observation. Indeed the red sky at night theory goes right back to the time of Jesus, for he said: 'When it is evening, ye say, it will be fair weather, for the sky is red. And in the morning, it will be foul weather today, for the sky is red and lowring' (Matthew XVI, 2–3).

Many of the weather lore traditions are linked to farming and the sea, and there are hundreds of sayings. In this book I have only included those which are in common use and which together make up a practical

weather guide. If they are consulted and used, then quite accurate predictions will be possible, and the more orthodox weather maps and long-range weather forecasts can safely be forgotten. An example of old lore being more accurate than modern forecasts occurred towards the end of the famous and enjoyable drought of 1976. Because of the parched conditions, one water authority decided, at great cost, to follow the advice of 'experts' and made a large East Anglian river flow backwards. Wise countrymen shook their heads commenting, 'The weather always equals itself out' and some men quoted the old saying:

Be it dry or be it wet,
The weather'll always pay its debt.

As soon as the pumps were installed it began to rain. The drought was over, and an exceptionally wet winter followed.

General Weather Rules

Although weather trends tend to be seasonal, there are many daily signs that give a guide to everyday weather. They are simple to observe and to interpret, and have been important to country people for many generations.

FINE

In addition to 'Red sky at night, shepherd's delight', there are other rhymes which confirm the importance of the colour of the sky at dusk and dawn.

Evening red and morning grey,
Two good signs for one fine day.
Evening grey and morning red,
Send the shepherd wet to bed.

Some of the early observers were more concerned with the sheep than the shepherd, however, and they said:

If the evening be grey and the morning red,
The lamb and ewe will go wet to bed.

In some rhymes the poor shepherd has been ousted altogether and replaced by the 'ploughman'. In many places, with the expansion of arable agriculture, this version is now much more appropriate. But whether involving shepherds, sheep, or ploughmen, a red sky at night does indicate a good day to come. It is important however that the red sky is not confused with an 'angry sky', for when the sunset is a fierce red, then stormy, squally weather is promised.

Dew in the night,
Next day will be bright

is also accurate, and a dry lawn on a summer morning is an ominous sign. Another reliable summer observation is:

Grey mists at dawn,
The day will be warm

which means the same as 'A summer fog is for fair weather'.

Some people insist that 'if you see a patch of blue sky in the early morning big enough to make a sailor a pair of trousers, the day will be fine'. I have not found this to be entirely accurate, probably because the 'bell bottoms' of modern-day sailors are smaller than those of ancient mariners and less blue sky is needed for their manufacture.

Rain before seven,
Fine before eleven

is another quite dependable rule, and a fine day can result from a bad start. This is also the case when the clouds break up before the wind, and it is true to say that:

A sun shiny shower,
Won't last half an hour.

As well as these rhymes, there are a number of signs that indicate good weather. Mirages promise good weather for a fortnight, and smoke rising vertically augurs well. For those who live within reasonable distance of a railway line, there is some confusion, for many people claim improved hearing means fine weather. They are probably wrong, however, for it is also said:

Sound travelling far and wide,
A stormy day will betide.

CHANGE

When good weather is about to break up, the signs are obvious – there are changes in the clouds, the wind, the temperature, and the visibility. The length of time in

which the change builds up also tells how long the new weather will last:

Long foretold, long past,
Short notice, soon past.

But change can be indicated in other more painful ways. People with bad feet feel their corns jumping, those who have rheumatism suffer from bad backs or niggling shoulders, and discomfort is felt where bones have been broken and long mended. I have broken an arm at cricket and my collar bone at football, and both joins sometimes become troublesome when cold rainy weather is on the way. A few people also experience tingling or drumming in the ears.

RAIN AND STORMS

In much the same way as a red sky at night is a good sign, so an anaemic sky prophesies rain:

If the sun goes pale to bed,
'Twill rain tomorrow it is said.

Unfortunately this, as well as most of the other warnings of rain, seems to be very reliable. Smoke falling instead of rising, 'white horses' out to sea, and ditches and drains smelling unpleasantly, are all bad signs, as is soot falling down the chimney. This is caused by the soot absorbing the increasing dampness in the atmosphere until it becomes heavy and falls into the grate.

Before the age of mass production and imports from Japan, the blacksmith's anvil acted as a reliable barometer, for if an anvil 'sweats' it is a promise of rain, and the larger the drops, then the heavier the fall. In winter, 'When the sun gets round into the wind', rain follows quite quickly. 'Three rimy frosts, and then it rains' is another saying, and the same is also said to apply to three fogs.

Once the rain has started to fall other rules apply:

Rain from the East,
Will last three days at least

and: 'If the rain comes down slanting, it will be everlasting.' A less reliable rule is: 'Wet Friday – wet Sunday.'

ICE AND SNOW

There are not many warnings about the approach of ice and snow, apart from the obvious fall in temperature, but when it is going to be cold it is true that fires seem to burn brighter in the grate. Our open fire is an exception to this rule. Usually, during cold weather and especially when it comes from the east, the fire smoulders and smokes, throwing out little heat. 'Black frost, long frost; hoar frost – three days and then rain' gives an indication of the length of the cold spell. It is also said that 'if snow hangs about [along ditches and hedgerows], then it is waiting for more.'

WIND

The wind which is most disliked by countrymen is the winter east wind, originating from the cold heart of Europe:

When the wind is in the East,
'Tis neither good for man or beast.

The east wind is said to be a 'lazy wind'; one that 'won't blow round you, but it blows straight through you'. However:

When the wind is in the West,
Then the wind is at its best.

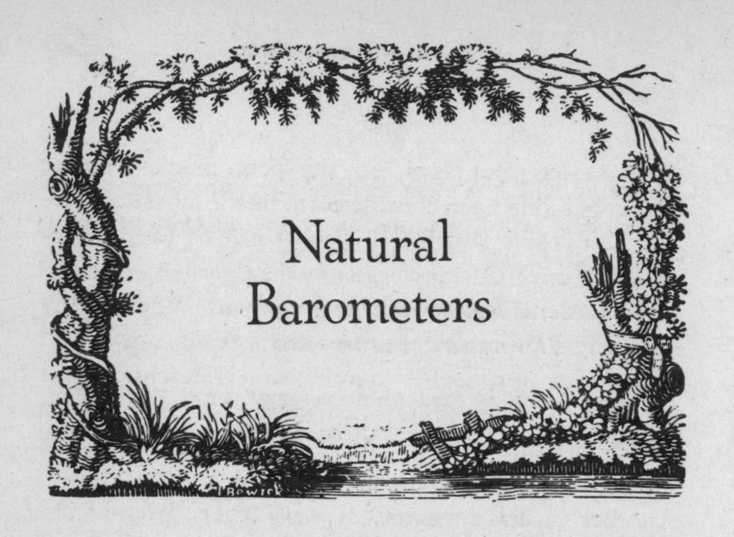

Natural Barometers

Barometers are always useful in weather prediction, and natural ones are particularly helpful. The two most common are pieces of seaweed and cones from fir trees. At one time, seaweed was always brought home by country people after a day at the sea, to hang in their porches or pantries, but now the custom seems to have lapsed. During fine weather seaweed shrivels and feels dry to the touch, but when rain threatens it expands and feels damp. Cones also give simple guidance. They open for good weather and close for bad. I have a Norwegian spruce cone as my barometer.

Fortunately for those who seldom go to the seaside, or who have no pine trees close at hand, there are many more aids, and on any country walk there are various flowers and trees which clearly show what weather can be expected.

FLOWERS

The scarlet pimpernel, or 'the poor man's weather glass', has long been considered to be the most accurate and responsive weather flower. It can be found anywhere, from roadside verge to urban garden – though most gardeners consider it to be a weed. When sunny weather is promised the tiny red flowers open, and when rain approaches they close. Dandelions and daisies also tend to shut when bad weather comes, and a slight variation of this can be found in Irish weather lore – 'when the dandelions and daisies close, it is dark'. Another garden aid is the ordinary onion, grown from a set or from seed. When a hard winter is promised, onions grow thick skins.

Several attractive flowers warn of cold weather, even after the advent of spring. In late April and early May there is often a spell of bitter wind which coincides with the flowering of the blackthorn – 'blackthorn winter'. The flowers of the blackthorn usually arrive before the leaves, and the bare black wood with its white florets looks cold and feels cold. However, it has its compensation, for sloes, the small dark fruits of the blackthorn, make excellent warming wine.

Many farming folk insist that you never get warm settled weather until the cowslips, or 'paigles', are finished. Some put the advent of warm weather even later:

You may shear your sheep,
When the elder blossoms peep

which is early June. A similar warning of late cold weather is given saying:

Ne'er cast a clout,
Before May is out.

Some say that this couplet refers to May the month, while others insist that it is 'may' the flower – the flower of the hawthorn tree. As a child it was recited to me as meaning the month, but there are other country people who remain adamant that it is the flower. If it is the flower, then the 'clout' can be left off a few days before the end of May the month.

TREES

Most people, whether they live in town or country, have heard a version of the common oak and ash prediction:

If the oak is out before the ash,
Then you'll only get a splash.
But if the ash beats the oak,
Then you can expect a soak.

Despite being so simple, many would-be weather experts manage to get muddled up and produce the wrong conclusion.

Trees can be used to make a forecast in other ways.

When they show the underside of their leaves, giving a much lighter appearance than usual, it is a sure sign of wet weather. The phenomenon is caused by the increasing moisture in the atmosphere which softens the stalks of the leaves, causing them to turn over. This is particularly noticeable with poplar, lime, sycamore, and lilac. Another sign of rain is if the wind makes a hollow sound among woodland or forest trees. When extra dry weather is likely, particularly in the autumn, trees may snap or crack. On one occasion I was walking through a wood in late summer when, for no apparent reason, a tree toppled over, making a noise like a burst of machine-gun fire: a long dry spell followed.

It is claimed by some that hard weather is indicated by an abundant crop of hips and haws – the fruits of the wild rose and the hawthorn. They claim that nature supplies the fruit as extra food to enable the birds to survive the winter. I believe that the reverse is true, and have noticed hard winters when there have been few hips and haws. This is nature reducing the population of birds and contributing towards a natural balance.

Occasionally, strange weather combinations produce bumper crops of walnuts, which are greeted thankfully by the owners, as the walnut tree is a most irregular cropper. Interestingly, the weather which produces walnuts also produces children, and in the past a good walnut year was always followed by many illegitimate children. In 1823 William Cobbett came across an old man who claimed that: 'A great nut year was a great bastard year . . . He said that he was sure that there were good grounds for it and he even cited instances in proof and mentioned one particular year, when there were four times as many bastards as ever had been born in the parish before; an effect which he ascribed solely to the crop of nuts the year before.' Modern methods of family planning now conceal the activities of young people during a glut of nuts, but parents who are worried about the activities of their daughters, are advised to lock them up whenever they see boughs laden with walnuts.

In much the same way it is also said that: 'When elder leaves are as big as a mouse's ear, then women are in season.' It is not quite clear whether this advice was said for the benefit of young men, or as a warning to anxious fathers and husbands.

ANIMALS, FISH AND CREEPING THINGS

With a minimum of knowledge about natural history, and with no stalking skill, it is possible to get quite an accurate weather forecast from the behaviour of an assortment of wild and domestic creatures, varying in size from cows to spiders.

For some reason, the two animals most commonly associated with rain are cats and dogs. Nobody seems to know why, yet 'It's raining cats and dogs' is in everyday parlance. Occasionally an odd character can be found who claims that the state of a cat's whiskers shows what weather to expect, for they can be stiff or droopy, depending on good weather or bad. My vet disagrees and informs me that a cat's whiskers say more about the health of the cat than they do about the state of the weather. However, it is true that cats become extremely frisky when high winds are likely.

Some dogs become very agitated at the approach of thunder, and they hear or feel the storm long before the tell-tale rumbles are heard by human ears. Our highly strung border collie will even jump up to hit the door handle with her paw and let herself into the house when storms threaten. Other creatures to be affected by thunder are rabbits, which all sit looking in one direction with their ears twitching, and toads, which hurry towards water. For some obscure reason eel fishermen catch more eels before a thunder storm.

It is a long established country belief that pigs can

see the wind. Consequently, when gales are on their way, pigs become very restless; they throw straw about with their snouts and rush around their sties.

Goats are also influenced by the wind. If a goat grazes with its head into the wind, it is going to be a fine day, but if it grazes with its tail into the wind, then

it will rain. If when observing goats, a 'billy' should look you up and down and sidle towards you with an evil expression on his face, and with his horns at the ready, get out of the field as quickly as possible.

Cows make particularly good weather indicators, as can be seen from the following rhyme:

When a cow tries to scratch its ear,
It means a shower is very near.
When it clumps its side with its tail,
Look out for thunder, lightning and hail.

In hilly country, when it is likely to stay fine the cows remain near the hilltops, but, if they are seen sniffing the air and then walking down hill towards the farm-yard, then rain or a storm will follow. Apart from the ability to smell storms, in summer they also warn of rain and thunder by stampeding with their tails in the air. This is because troublesome flies become more active during the warm, humid build-up to a storm, particularly the warble fly, which lays its eggs in the cow's hide. It must be a painful experience, for even on a cool day, if a high-pitched buzz is made behind a cow, it will shoot its tail into the air and run away. Indeed, whenever we drive our cows to the farm for milking, and they start to dawdle and graze on the roadside verges, we can hurry them along by imitating the buzz of the warble fly.

In colder weather cows will lie down and chew the cud when rain approaches, or they will huddle together in a field corner, standing with their tails to the wind.

Although they have wings, bats are mammals, and when they fly out after flies in the late evening, a fine day should follow. If a long dry spell is likely, moles dig holes as usual, but they build no hills on the surface, and their presence can be detected simply by the slight outlines of their tunnels. Fish, too, can forecast hot weather, for if temperatures are likely to be high they keep to the cooler, lower water, and are reluctant to take a hook, but when rain is imminent they gambol about near the surface, causing ripples and splashes.

One of the most accurate ways of foretelling change has become much less popular over recent years. It involves sharing your house with mice. When a change is in the air mice become more active. They move from room to room, or from skirting board to skirting board, and they can be heard squeaking.

Another sign of change, and probably a promise of very heavy rain with floods, is the sight of hares moving from low ground to high ground.

The most common animal indications of rain are: rabbits out at strange times of the day, deer feeding early, donkeys braying more loudly than usual, and cats washing fastidiously behind their ears. In the few

places where sprays and drainage experts have been controlled, frogs croak with great enthusiasm and large numbers of toads appear in the evening when rain threatens. It is of interest to note that in winter, if there is a long cold spell, pregnant sheep will hang on to their lambs. But when it rains, bringing with it a higher temperature, they will all lamb at once: an example of nature and the weather working together for the benefit of the lambs.

Fortunately spiders are abundant throughout the country, and they forecast several types of weather. When the day is going to be hot and dry they spin long webs but, when bad weather threatens, the webs are short or they disappear altogether. Some people even claim to have seen spiders breaking their own webs before bad weather. Before windy weather the main strands are tightened, and the spiders behave rather like campers adjusting the guy ropes to their tents. During some summer days it actually seems to be raining spiders, when thousands of small spiders, each with its own short strand of gossamer thread, drift by in the breeze: this is another good weather sign.

Gnats are not as attractive as birds or bees, but they are useful weather forecasters in both winter and summer. If they play up and down in a swarm out in the open, a sunny day is forecast. If they remain in the shade, showers are likely.

BIRDS AND BEES

Although the British are often described as a nation of animal lovers, it would be more accurate to describe them as a nation of bird lovers, for each year thousands of bird tables and nest boxes are erected, food for wild birds is sold ready mixed in packets, and more and more people describe themselves as bird watchers (if they do it for fun), or ornithologists (if they take themselves more seriously than the birds they watch). This is a good thing, for birds, together with bees and gnats, give accurate weather guidance. They also make natural weather forecasting pleasant, giving memories

of swallows flying high, waxwings eating winter berries, and honey bees returning to their hive.

Many birds can indicate fine weather. When the robin sings on rooftops and at the top of trees it is a good sign, but if it gets down lower, then be prepared for rain. Blackbirds singing from the treetops also promise a fine day, but again, if they change to the bottom branches, beware. If a blackbird sings with its tail straight down, that is also a warning, for it is 'waiting to shoot the water off'. Geese flying out to sea and free-range hens scrapping contentedly are other signs of good weather. Pheasants, too, are dependable: if they go up to roost late in the evening, and are up with the sun the next morning, a good day is promised, but if they are early to perch, or late down to feed, the weather will break up.

The skylark is another bird that people usually associate with good weather. They claim that if it hovers and glides during its descent, the weather will remain fine, but if it drops straight down to the ground, it will rain.

When a missel thrush, from the very top of a tree, sings into the wind, it is a bad sign, and this habit has led to the bird acquiring the name 'storm cock' or 'storm thrush'. Even the song signifies rain, for the

bird sings: 'More wet, more wet.' The green woodpecker has gained a similar reputation, and because its laughing call is often heard before rain, in some areas it is known as 'the rain bird'. In addition, as rain

approaches, ducks and geese quack and cackle more loudly, and the cry of the screech owl becomes harsher; hens huddle together outside their houses, and:

If the cock goes crowing to bed,
He'll certainly rise with a watery head.

Another old saying warns that 'the hoarse crow croaks before rain'.

At one time, the sight of seagulls flying inland was a certain sign of rain. However, with the growth of inland sewerage works and the development of large reservoirs in once peaceful and fertile agricultural valleys, some seagulls have now changed their habits and live completcly away from the sea. Consequently the old rule continues to apply only where there are no reservoirs and where the bucket lavatory still reigns supreme. Mention of bucket lavatories suggests flies, and they can also anticipate the approach of rain. This is particularly true when they settle in large numbers on walls and ceilings. Some people also quote:

If a fly lands on your nose, swat it till it goes,
If the fly then lands again, it will bring back heavy rain.

In coastal areas the sighting of a storm petrel is a sign of bad weather. 'Mother Carey's Chicken' is usually a bird of the open ocean, only travelling to remote cliffs or rocky islands to breed. It is an attractive little bird, but if it is sighted near land, away from its breeding sites, it has probably been driven inshore by bad weather, which will soon follow.

Very cold weather and long hard winters are shown by the early arrival of winter migrants such as ducks, geese and swans, together with much larger than usual flocks of fieldfares and redwings. The birds also travel further south than in an average winter. The arrival of the unmistakable waxwing from Scandinavia can also mean severe weather.

Although 'one swallow doesn't make a summer', swallows and swifts are welcome and attractive summer visitors. When they fly high, following the insects on which they feed, it is safe to say that the day will be fine. Sometimes they can attain such heights that they appear just as faint fleeting specks in the clear blue sky. Occasionally the rule is broken, for sometimes they fly high during thundery weather; but then the heavy black clouds are obvious, and the swallows cannot really be blamed if the umbrella is still left indoors.

For those who live near rookeries, these noisy sociable birds make excellent forecasters for all weathers. When rooks fly from their nests and fly straight, umbrellas and raincoats can be left at home all

day, although the observer must be up at dawn to see them go. Should they twist and turn on leaving their nests, rough weather is approaching. However, if gales are on the way, they stay by their nests, screaming raucously. When they are late leaving the rookery in

the morning and then hang about, feeding on the roadside and in village streets, it will rain. The twig nests are also important, as it is said that high nests mean a good summer. Low nests suggest the destruction of the old nest sites during winter gales, and so building lower, on more secure branches, can indicate a particular weather pattern.

Cuckoos feature in much poetry and folklore, and

most children learn a variation of the verse that was taught to me as a child:

The cuckoo comes in April,
He sings his song in May,
In the middle of June he changes his tune,
And in July he flies away.

But strangely, cuckoos feature in very little weather lore, although it is said that if a cuckoo can be penned up in an enclosure of hedges and trees, to prevent it from

flying away, then summer will never end. As no cuckoo has yet been trapped in this way the statement can be neither proved nor disproved.

Anybody who has a hive of bees can regard the

inhabitants as reliable weather guides, and there are many rhymes and rules concerning bees:

A swarm of bees in May is worth a load of hay,
A swarm of bees in June is worth a silver spoon,
A swarm of bees in July is not worth a fly.

This is commonly said and reflects both the weather and the amount of money that the bee-keeper will get

for his honey during a good and a bad season. Bees thrive on the sun and the verse is true which says:

When the bees crowd out of their hive,
The weather makes it good to be alive.

They also hate rain and storms, which makes the final two lines equally appropriate:

When the bees crowd into their hive again,
It is a sign of thunder and of rain.

As a rule 'bees will not swarm before a storm', and if thunder is near, the good bee-keeper will keep away from his hive as the residents become very angry, crowding back into the hive and stinging anybody who gets in their way.

CELESTIAL BODIES

The other main elements of weather forecasting involve the clouds, thunder and lightning, rainbows, the stars and the moon. They can all be used in conjunction with the various other rules to build up an accurate weather picture.

CLOUDS

These are obviously a great influence, as they dictate sun and shade, rain and snow, and they can also indicate

wind. The most pleasing clouds of all are those which resemble small white patches of fluff, for they are normally associated with fine weather.

If woolly fleeces strew the heavenly way,
Be sure no rain disturbs the summer day.

When cirrus clouds appear, high up and wind-whipped, like mares' tails, they foretell the arrival of wind:

Mackerel sky and mares' tails,
Make tall ships carry low sails.

Much weather lore associated with clouds has a nautical link, as weather is a dominant factor in the

lives of sailors as well as countrymen. Because of this there even seems to be one rhyme composed specifically for sailors and farm workers:

When clouds look as if scratched by a hen,
Get ready to reef your topsails then.

A mackerel sky, as its name suggests, is mottled like the fish, and when the sky gradually clouds over with this mackerel pattern, fine weather is on the way out. There are many rhymes and verses about the mackerel sky: 'Mackerel sky – rain is nigh' is one, but usually the rain will not be heavy:

Mackerel sky, mackerel sky,
Never long wet, never long dry.

Longer periods of rain can be forecast by the appearance of 'the ark'. It is a thin strip of high feathery

cloud that forms an archway across the sky. Sometimes the ark is not quite complete and ends in a point of elongated cloud. It is said that: 'When the ark is out, rain is about', but there are still some old countrymen who claim that to bring rain it must be pointing in a certain direction:

When the ark is out,
North and South,
In the rain's mouth.

If the clouds appear to build up in large blocks, showery weather is promised, for:

When clouds appear like rocks and towers,
The earth will be refreshed by frequent showers.

Or:

A rain-topped cloud with flattened base,
Carries rain drops on its face.

When the clouds build up to form a towering black anvil, a thunder storm is almost certain.

THUNDER

Thunder at unusual times of the year seems to affect the weather later on: 'Winter thunder, rich man's food, poor man's hunger', and 'Winter thunder, summer hunger', both indicate a bad year, while it is commonly said that in an English summer you get 'three hot days and a thunder storm'. Many children are frightened by thunder and some claim that 'it is God moving his furniture about'. A large number of older people, particularly women, retain their childhood fear, and seek shelter beneath tables, or under the stairs, to avoid being struck by lightning. Mistletoe hanging in a room is said to afford good protection and, in the past, it was not unusual for country cottages to grow a house leek on the roof to keep the lightning away.

If out in the open when a storm breaks it is claimed that oak trees give safe shelter – although I would not be willing to put the belief to the test myself – but ashes and elms are considered to be very dangerous:

Avoid an ash – it courts a flash.

THE MOON

Whenever there is an open sky at night, the stars and moon give valuable forecasting assistance. A 'watery moon', when the moon shines weakly through haze, indicates rain, and one of the most reliable weather rules associates the halo that sometimes appears round the moon with wet weather.

Near ring far rain,
Far ring near rain.

'When the moon is on its back, it holds the rain in its lap' means a fine day, according to some, as it does when 'the old moon is in the young moon's arms'.

Then the outline of the old moon can be clearly seen as a dark shadow above the new moon. A clear moon

means a fine day, while in winter it forecasts frost: 'Clear moon – frost soon.' The colour of the moon is also important in determining the weather:

Pale moon does rain,
Red moon does blow,
White moon does neither rain nor snow.

It is also believed that 'the weather that comes in with the moon will stay like it for a month.' Some gardeners will only plant seeds when the moon is on the wane.

STARS

To the casual observer, one of the surprising facts about the night sky is that on some nights there appear to be more stars than on others, and that the brilliance of the stars themselves also seems to vary.

In the past, wise countrymen have noticed these changes and linked them to certain types of weather. Like the moon, when stars appear to be particularly bright, it means good weather, and the fewer there are in number the better. But when they twinkle it is a sign of wind.

There are also occasions when the sky seems to be packed with stars, an astronomer's paradise. This is a bad sign:

When the stars begin to huddle,
The earth will soon become a puddle.

From their position in the sky the stars show the season, and the easiest formation to pick out is the Plough, known by many old country people as 'Dick and his wagon'. When the handle is pointed upwards, Dick is taking his wagon up the hill and it is likely to be fine. When he is taking his wagon down the hill it is more likely to be wet.

RAINBOWS

Apart from the Biblical promise associated with rainbows, a rainbow in the morning or evening can show whether it is going to be a good or a bad day:

A rainbow at night,
Fair weather in sight.
A rainbow at morn,
Fair weather all gorn.

Occasionally double rainbows can be seen, and they show that the weather is clearing. If such bows are seen beneath you, rather than above, proceed with care; you are probably on the edge of the Victoria Falls, a beautiful and awesome sight, but not good for walking over.

Making a Forecast

LONG-RANGE WEATHER FORECASTING

If the sayings and conditions so far recorded have been remembered, it should already be possible to make quite accurate long-range weather forecasts. One or two additional guides include: 'A good winter brings a good summer' and 'Early thunder, early spring'. In *Lorna Doone*, John Ridd quotes a saying from the West Country:

A foot deep of rain,
Will kill hay and grain;
But three feet of snow,
Will make them grow mo'.

When horses were used on the farms it was also said that if they grew thicker coats than usual, it was a sign

of a cold winter, and during the same farming period, if a wooden cart had a squeaking axle it was said to show fine weather.

TECHNOLOGICAL WEATHER LORE

Just as a lot of weather lore developed in the past, new lore is still being created and modified to cope with changing life styles and technical advance. One old countryman now likens 'the ark' to an elongated Concorde, while an old woman claims that if aeroplanes are seen flying high, it will be a fine day. This is not quite as senseless as it seems, for during bad weather no high-flying planes with their white vapour trails can be seen because of cloud.

Television reception appears to be better in fine weather, for when storms are in the offing our television set picks up gabbling Frenchmen. If thunder is in the air, those who weld with D.C. welders experience problems, and in some windy spring weather the smell of T.C.P. can sometimes be detected in the air. This does not indicate bad weather, but bad farmers spraying the whole countryside instead of waiting for still weather and covering only their fields.

In arable areas the farmers are also responsible for another interesting development. When a field is combined, much of the threshed straw is burnt before the rest of the harvesting is finished, sending clouds of smoke and smuts into the atmosphere. Often, as soon as the amount of stubble burning reaches a certain level, the weather changes, the sun becomes hazy and rain falls. This could be based on the fact that moisture needs a speck of dust around which to form. Consequently, stubble burning provides plenty of ideal

particles and so by excessive burning farmers could be hindering the completion of their own harvests:

In September after burning stubble,
Ponds and streams begin to bubble

and a new piece of weather lore is born.

THE OUTLOOK FOR TOMORROW

Much weather lore is local, related to specific hills, lakes or forests. In the Vale of Evesham it is said that:

When Bredon Hill puts on his hat,
Ye men of the vale beware of that.

Similarly in Devon there is Franklin's frost, which strikes on 19, 20 or 21 May.

Because local conditions vary it must also be admitted that some weather lore is contradictory. Some people say that the moon on its back is a sign of rain, and there is even disagreement about the ash coming out before the oak, as there is a rhyme which proclaims:

If the ash before the oak,
We shall have a summer of dust and smoke.

These contradictions can be explained by the fact that situations vary according to soil, shelter and the locality, and so a different set of rules can apply in some areas. It also means that, like professional weathermen, countrymen can always find good excuses if their forecasts go wrong.

Nevertheless, in general, taking all the weather rules together, an accurate picture can usually be obtained, and if contradictory rules appear at the same time, common sense can establish the best forecast. Recently, around the farm, there were signs of both sun and rain, but by following the most obvious it was possible to predict fine weather. The swallows were high, there was a red sky at night, the stars were clear and widely spaced, 'the young moon had the old moon in its arms', my spruce cone was open, and the 'may' was out, so I could cast a clout. However, the rooks never ventured far from the near-by rookery, and their flight was erratic. Hot weather came, and it is likely that the behaviour of the rooks was caused by the fact that they

had young in their nests, and that the sun and wind actually encouraged them to tumble and glide in the warm turbulent air, for sheer enjoyment.

If common sense, therefore, is applied to country weather lore it should be possible to choose the right days for holidays, gardening and visits to the Test Match. Barometers can be consulted to aid accuracy, but the more orthodox forecasts and forecasters can be abandoned almost completely.

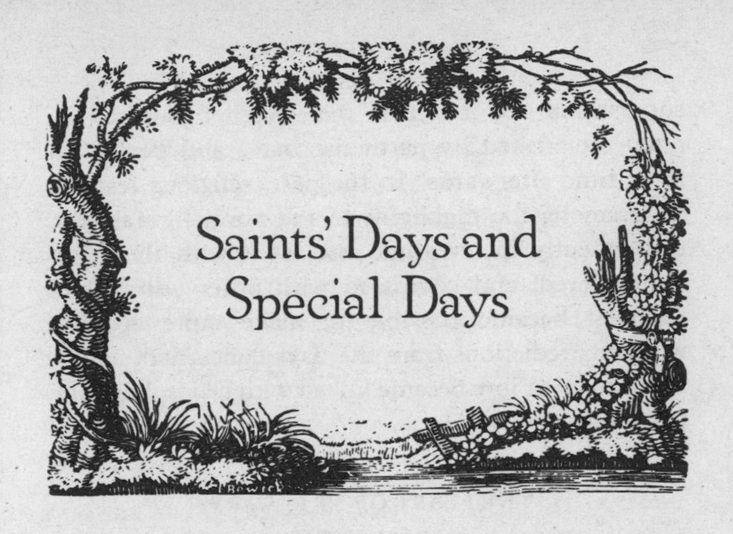

Saints' Days and Special Days

There are a number of important days throughout the year which are thought to influence the weather, and

which can aid forecasting, and many of them are linked to days in the Church calendar. They are not based on superstition, for it is true that weather patterns are often established by particular dates and persist for some time afterwards. In the past, religious festivals were among the highlights of the normal rural year, consequently the weather associated with them was remembered, and compared with other years. As a result it became possible to make quite accurate weather predictions from the days concerned, and so much weather lore became linked with feasts, holidays, and saints' days.

THE FEAST OF ST HILARY, 13 JANUARY

This is often considered to be the coldest day of the year.

ST PAUL'S DAY, 25 JANUARY

If St Paul's Day be fair and clear,
Then it betides a happy year.

Some versions go even further, suggesting that if the day is fine there will be good harvests; if there is rain and snow, there will be scarcity, and if there are clouds and mists there will be pestilence, high winds and war. It is the day that commemorates St Paul's miraculous conversion.

CANDLEMAS DAY,
2 FEBRUARY

If Candlemas be fair and bright,
Winter'll have another flight.
But if Candlemas Day be clouds and rain,
Winter is gone and will not come again.

There are many versions of this common Candlemas rhyme. It is also said that on Candlemas Day the livestock farmer must still have half his straw and half his hay, emphasizing that although February has been reached, winter has a long way to go.

Candlemas Day itself is the day when the child Jesus was taken to the temple by his mother and the old man Simeon recognized him as 'a light to lighten the

Gentiles and the glory of thy people Israel' (Luke II, 22–38). The day gets its name from the fact that by the middle of the fifth century it was celebrated with lighted candles.

LENT

A dry Lent means a fertile year.

EASTER DAY

Easter in snow, Christmas in mud,
Christmas in snow, Easter in mud.

It is also said that:

If it rains on Easter Day,
There shall be good grass but very bad hay.

In other words, haytime will be wet, which means a wet June.

Despite the fact that Easter is a movable feast, these sayings have been handed down as definite weather traditions as they have become firmly associated with the festival.

ST VITUS'S DAY, 15 JUNE

If St Vitus Day be rainy weather,
It will rain for thirty days together.

Evidently St Vitus is not only remembered for the sickly dance named after him, but also for rain on his day.

ST MARY'S DAY, 2 JULY

It is said that if it rains on St Mary's Day, it will rain for a month.

ST SWITHIN'S DAY, 15 JULY

Oh St Swithin if thou'll be fair,
For forty days shall rain nae mair,
But if St Swithin's thou be wet,
For forty days it raineth yet.

St Swithin's is undoubtedly the most famous and the most notorious weather day of the year, for it is the popular belief that if it rains on St Swithin's it will rain every day for the next forty days. Even now farmers take this date very seriously, for a dry St Swithin's can mean a less worrying harvest. On our farm we certainly take notice of the weather on 15 July and view rain on that day with anxiety.

St Swithin himself was a humble monk who suggested that on his death his body should be buried where the rain would fall on him. This was done in 862 A.D., but in 971 A.D. an attempt was made to move his remains to inside Winchester Cathedral, which was considered to be far more appropriate for a man of his godliness. It is said that his spirit was so outraged that it made the rain fall for the next forty days, until the monks gave up and returned to more fruitful labours.

ST BARTHOLOMEW'S DAY, 24 AUGUST

If St Bartholomew's Day be fair and clear,
Then a prosperous autumn comes that year.

If the weather is settled on this day, a fine autumn is promised.

ST MICHAEL AND GALLUS, 29 SEPTEMBER AND 16 OCTOBER

If it does not rain on St Michael and Gallus,
The following spring will be dry and propitious.

This is a piece of weather lore for the more literate. Those who use the word claim that 'propitious' means 'inclined to show favour' or 'of good omen'.

ST LUKE'S DAY, 18 OCTOBER

St Luke's Day is the other really important day for farmers, as the days surrounding it make up 'St Luke's Little Summer'. It is a period of time that father always refers to during a wet harvest, as it usually offers a few

days' respite, so that a late harvest can be finished. I have only known one season when crops have been lost completely because of rain in August, September and October.

ST SIMON AND ST JUDE, 28 OCTOBER

Like St Swithin's, this day has a reputation for bad weather. It is a time when the weather breaks up, gales begin, and sailors make for harbour at the first sign of wind. It can also mark the end of St Luke's Little Summer.

ST MARTIN'S DAY, 11 NOVEMBER

St Martin, too, has a 'little summer' named after him. It is not really appropriate, however, as St Martin is the patron saint of reformed drunkards, and so it ought to be a time of heavy rain with abundant fresh water.

ST THOMAS'S DAY, 21 DECEMBER

St Thomas grey, St Thomas grey,
The longest night and the shortest day.

CHRISTMAS DAY

Sun through the apple trees on Christmas day,
Means a fine crop is on the way.

According to tradition and observation it is important to have a fine Christmas Day to ensure a good spring and few late frosts. Frosts 'cut' apple blossom and ruin crops, so a sunny Christmas means a frost-free May and a good autumn.

The Calendar of Weather

January

The type of weather experienced is obviously dominated by the time of year and the position of the sun. Yet it is also influenced by weather patterns, and by remembering the storms or heat waves of one month or one season, it is possible to guess quite accurately the weather of the next.

January is an unreliable month. It can be very cold, or surprisingly warm, and can be summed up by the old proverb: 'Winter weather and women's thoughts often change.'

If the grass grows in January, it grows the worse
for all the year.

A January spring is worth nothing.

If the birds begin to sing in January,
Frosts are on the way.

March in January,
January in March.

February

February is a damp month, not because of high rainfall, but because of a low evaporation rate. It is often the month of the most intense cold, as the thermometer falls and the crimson sun sets in an open sky. It can be a time of burst pipes, and, in a good year, of skating.

*February fill dyke, black or white.**

February makes a bridge and March breaks it.†

As the days lengthen,
So the cold strengthens.

---

*This means that regardless of rain or snow the ditches will usually fill during February.

†A bridge of ice which thaws in March.

March

It is usually claimed that spring starts on 21 March, but in fact 'it is not spring until you can put a foot down on twelve daisies'. Daisies are among the first flowers to respond to the warming sun and the rising sap, but many springs do not really start until April or May. When spring is eventually with us even my small 5½-sized feet can come down on as many as thirty daisies each, but in areas where too much chemical spray is used, then perhaps the number for the first day of spring should be reduced to three.

March is usually a very varied month and the sensible traveller will be prepared for anything.

March many weathers

If March comes in like a lion,
It goes out like a lamb.
If it comes in like a lamb,
It goes out like a lion.

*March borrows its last three days from April.**

As many mists in March as there are frosts in May.

---

*Like scientific weather forecasters, country weather experts have their disagreements and some claim that March takes ten days from April. Whichever is correct the weather at the end of March and the beginning of April is usually similar.

April

April is most famous for its showers, which are often at their heaviest on the last Saturday of the month, the first day of the cricket season. It is also a month which gives an indication of the weather to come later on in the summer.

Thunder in April,
*Floods in May.**

A cold April and a full barn.

When April blows his horn,
'Tis good for hay and corn.†

---

*In some areas the saying is: *Thunder in March, Floods in May.*

†The 'horn' is a reference to thunder. If good hay and corn are promised, it also means a fine summer.

May

The merry month of May is my favourite month. It is the month when birdsong, flowers, blossom, and new foliage are all at their best. It is also a month of variable temperatures and rainfall. Traditionally a cold May is better both for people and for harvest.

March winds and April showers,
Bring forth May flowers.

A wet May,
*Brings a good load of hay.**

A cold May and a windy,
Makes a fat barn and a findy [*good weight*].

A hot May makes a fat churchyard.†

---

*This means plenty of sun in June.

†It does seem that a warm May makes extra work for florists, undertakers, and crematorium attendants.

June

Flaming June has nothing whatsoever to do with crematoriums, but as a rule June is a good month, with spells of settled weather. Warm June evenings are ideal for watching badger and fox cubs, and it is a month when young birds and animals can be seen at their best. On the coast it is also the time to see razorbills, guillemots, and puffins at close quarters.

Calm weather in June,
*Sets the corn in tune.**

A dry May and dripping June,
Brings everything in tune.

A leak in June,
*Sets the corn in tune.**

---

*Another example of country weather forecasters not being in complete agreement.

July, August & September

There is little reliable weather lore concerning these months, partly because the first two often contain very variable weather, being the two wettest months of the year, but also because nothing much rhymes with July or August, and in any case the full quota of 'flowers' and 'showers' helps make up the verses for earlier months. The only useful word to rhyme with September is 'remember', but that is over-used later, in November, for 'Remember, remember, the Fifth of November'. I like September, settled weather is often the rule, and it is a month of gentle change when:

The squirrel gloats on his accomplished hoard,
The ants have burried their gaines with ripe grain,
And honey bees have stored
The sweets of summer in their luscious cells,
The swallows have winged across the main!

THOMAS HOOD

October

This is usually a month of falling leaves and falling temperatures, when the first black ice of the winter gives plenty of work to garages and insurance companies. A large number of foggy days in October indicates a hard winter.

If the leaves wither on their branches in late autumn, instead of beginning to fall in October as normal, an extra cold winter is due. This also confirms the 'ice and duck' rhyme of November, for early frosts often bring the autumn leaves down in October and November. Consequently when this does not occur, hard weather can follow.

November

Just as fog in October suggests a cold winter, so a cold November signifies a mild winter. All in all it is a dismal month, best summed up in the words of Thomas Hood:

No warmth, no cheerfulness, no healthful ease,
No comfortable feel in any member –
No shade, no shine, no butterflies, no bees,
No fruits, no flowers, no leaves, no birds, –
*November!**

Ice in November to bear a duck,
The rest of the Winter'll be slush and muck.†

---

*This is not altogether true, for I have picked good blackberries in November.

†This is often true, however, and was confirmed regularly by our farmyard muscovy ducks, before the wild foxes ate them. Consequently the best advice to any car-travelling-commuter who wants a reliable long-range-winter-weather-forecast is, 'first find your pond and then your duck'. But when confirming this item of weather lore it is important not to contravene motoring law, for driving without due care and attention, by watching ducks instead of cars, is a punishable offence.

December

December has very little weather lore connected with it. It is interesting to note, however, that despite containing the season of goodwill, 'a green Christmas [like a warm May] means a fat churchyard,' and, again, it seems to be based on fact.

Index

Index

COLIN ELFORD

A YEAR IN THE WOODS

'In the dark, with the rifle on my back, I make my way up the steep track leaning into the hill. This is my time, and I love it, free from people, just me in the woods.'

Colin Elford spends his days alone – alone but for the deer, the squirrels, the rabbits, the birds and the many other creatures inhabiting the woods.

From the crisp coldness of January, through the excitement of spring and the warmth of summer, and back into the dampness of the autumn and the chill winds of winter, we accompany the forest ranger as he goes about his work – stalking in the early-morning darkness, putting an injured fallow buck out of its misery, watching stoats kill a hare, observing owls, and simply enjoying the outdoors.

A Year in the Woods is an invigorating look at nature through a forest ranger's eyes and an enthralling journey deep into the woods.

'This is nature for real . . . Elford describes woodland wonders in short paragraphs of luminous intensity' Val Hennessy, *Daily Mail*

'There is an almost hallucinatory clarity to some of Elford's writing . . . even his briefest dispatches possess such unsentimental intimacy' *Observer*

'Wonderful descriptions and sentences which are so profound they demand a second reading . . . a curiously comforting book which reminds us that however harsh the natural world, it will have its kinder moments' *Daily Express*

CHRIS YATES

HOW TO FISH

How to Fish is an unabashed, unashamed celebration of the joys of fishing. It is about contentment, calm and solitude, rivers and river banks, losing track of time and, of course, the fish themselves. For those who already enjoy fishing it is a love letter to their art and for those who don't – yet! – it is an insight into a life spent getting up at the crack of dawn and, armed with rod and line, heading for water…

'Will delight anyone who has ever fished and anyone else who has wondered what the appeal of fishing might be' *The Times*

'There can be few who have explained so elegantly both the Zen-like trance of the serious fisherman and the all-too-active comedy of preparation, expedition and return' *Guardian*

'It's a book about time, how to spend time and lose it and how to find some kind of peace at the end of it all. A wonderfully elegiac book that winds gently along with the quiet majesty of an old, old river' *Big Issue*

CHRIS YATES

ON FISHING AT SEA

'Though most of my life seems to have been spent on the banks of lakes and rivers, I have always been drawn to the sea …'

Through twenty-two casts, Britain's best-known freshwater fisherman quits land in favour of the sea. There, he discovers the many pleasures of the coast: wild shores, unpredictable waves, the violent collision of the elements and, of course, fish that glisten and dart beneath a never-still surface.

From childhood remembrances of saltwater escapades to more recent discoveries, Chris Yates brings the sea and its many wonders to scintillating life.

'There is more beautiful, securely thought, elegantly expressed and subtly responsive writing in this book than in volume after volume that make far greater claims to "literature"' *Daily Telegraph*

'One day Chris Yates's books will be available on prescription from your doctor, as an antidote to modern-day ailments with more power to soothe and subdue than any tranquillizer' Books of the Year, *Angler's Mail*

'There's something beautiful in the way Yates can evoke a shared love of the shore just by describing how a change in the wind or light makes it smell and look different' *Metro*

ROGER DEAKIN

WILDWOOD

'Wildwood is about the element wood, as it exists in nature, in our souls, in our culture and our lives'

Roger Deakin's unmatched exploration of our relationship with trees is autobiography, history, traveller's tale and incisive work of natural history. It will take you into the heart of the woods, where we go 'to grow, learn and change'.

'Enchanting, very funny, every page carries a fascinating nugget. Should serve to make us appreciate more keenly all that we have here on earth … one of the greatest of all Nature writers' *Mail on Sunday*

'Wonderful. Full of delight and joy and wisdom' *Sunday Telegraph*

'Extraordinary … some of the finest naturalist writing for many years. It is fitting that Deakin's last book should be his masterpiece' *Independent*

ROGER DEAKIN

NOTES FROM WALNUT TREE FARM

For the last six years of his life, Roger Deakin kept notebooks in which he wrote his daily thoughts, impressions, feelings and observations about and around his home, Walnut Tree Farm. Collected here are the very best of these writings, capturing his extraordinary, restless curiosity about nature as well as his impressions of our changing world.

'Marvellous, wonderful, lovely, remarkable . . . to be read and reread and treasured' Elizabeth Jane Howard, *Daily Mail*

'Very funny, sharp-eyed. To look at the world through Deakin's eyes was to see somewhere that was more wonderful than it often appears' *Sunday Telegraph*

'Thoughtful and invigorating, full of humour, timeless . . . will take its place among the classics of Nature diaries . . . to be read alongside Frances Kilvert, Gilbert White, and Dorothy Wordsworth'
Mail on Sunday

TOM HODGKINSON

HOW TO BE FREE

MODERN LIFE IS ABSURD. HOW CAN WE BE FREE?

If you've ever wondered why you bother to go to work, or why so much of consumer culture is crap, then this book is for you. Looking to history, literature and philosophy for inspiration, Tom Hodgkinson provides a joyful blueprint for a simpler and freer way of life. Filled with practical tips as well as inspiring reflections, here you can learn how to throw off the shackles of anxiety, bureaucracy, debt, governments, housework, supermarkets, waste and much else besides. Are you ready to be free? Read this book and find out.

'One of the most provocatively entertaining, creatively subversive and, frankly, essential manifestoes of this or any moment' *Time Out*

'As a follow-up to his charming *How to be Idle*, Tom Hodgkinson offers nothing less than a manifesto of resistance to the modern world' *Guardian*

'An inspiring collection for those of us who yearn to cast off our corporate and consumer shackles' *Psychologies*

TOM HODGKINSON

HOW TO BE IDLE

Society today extols the virtues of efficiency and frowns upon laziness, but as Oscar Wilde once said: doing nothing is hard work. As modern life grows more and more demanding, the loafers of this world can feel the odds stacking against them. But help is at hand! From Tom Hodgkinson, editor of *The Idler,* comes an antidote to the work-obsessed culture that puts so many obstacles between ourselves and our dreams. Learn how to reclaim your right to sleep in, skive off, lunch at leisure, have a hangover and take time out; and how to let the day slip past you in the best possible way. The message is clear: in so doing, you'll be taking control of your life.

Are you ready to be idle? Read this book and find out.

'A superb, life-enhancing celebration of idleness. A book to be enjoyed at leisure and to change lives' *Sunday Times*

'So stuffed with wisdom and so studded with good jokes that I raced through it like a speed freak' *Independent on Sunday* read more 

JAY GRIFFITHS

WILD: AN ELEMENTAL JOURNEY

'I took seven years over this work, spent all I had, my time, money and energy. Part of the journey was a green riot and part a deathly bleakness. I got ill, I got well. I went to the shamans in the Amazon for their hallucinogenic medicine and went to the freedom fighters of West Papua to sing my head off in their highlands. I met cannibals infinitely kinder than the murderous missionaries who evangelize them. I anchored a boat to an iceberg where polar bears slept; ate witchetty grubs and visited sea gypsies. I found a paradox of wildness in the glinting softness of its charisma, for what is savage is in the deepest sense gentle and what is wild is kind. In the end – a strangely sweet result – I came to a wild home…'

'Wholly original, undefinable, untameable, profound and extraordinary' *Observer*

'Utterly compelling, easily the best travel book that I have read in the last ten years' *Guardian*

'Incandescent, exhilarating, sensuous, cocky, magnificent, explosive. A raging oratorio' Richard Mabey, *The Times*
